はしがき

昨今、特に我国の少子化現象に伴ってのみならず、空前と言っても過言ではないぐらいのペットブームになっている。それに伴いペット関連産業も数千億の市場性があるといわれている。しかし、人間と同じ動物で『生命』というものがありながら、物として扱われてしまっているのが現実です。そして、それに携わる我々人間界は、ペットが家族の一員化してしまっているにも拘らず、例えば、獣医の医療においては、医療費格差があったり、(物である以上は仕方ないが)ペット屋で購入時も、価格格差があったり、不良品や「偽血統書を掴まされた」とか、道路で車に引かれて放置されていたりとか、飼い主の身勝手により捨てられたり、虐待・狂牛病・野畜による被害等挙げればきりが無いほど我々人間が解決してやらねばならない問題がいっぱいある。そこで、これら諸問題を以下の方法で現状と照らし合わせながら解決する。それらを、〔保険やクレジット〕を絡めることで、総体的に管理し情報を調整・交換し、提供して行くのを目的とする。

牧野真一

Preface

Our country is a pet boom in connection with the decrease-in-the-birthrate phenomenon.
This pet correlative industry also has hundreds of billions of marketability.
However, as for a pet, it is actual for it to be living and to be treated as a thing for the same animal as man.
A pet also considers as a family's member and has a medical-expenses gap in a veterinarian's medical treatment, and if it is a thing at all, although it is unavoidable, there is also a price gap at the time of purchase in a pet store, there are also inferior goods and a fake pedigree, and a human-beings [us who are engaged in it] community is pulled to a car in the road, and is left.
These are the problems which it was thrown away at will [an owner] more and man has to solve.
then, insurance be involved in order to solve these many problems -- it manages on the whole and adjustment and exchange are carried out for information

Shin-ichi Makino

目　次

１，各種サービスシステムの登録

動物の個々や品種により登録料は異なり、補償保険料等も含まれる。

⑴　一部の行政においては、犬に限定して行われているが、窓口を限定して、ペット屋で購入時することにより品質保証にもなる。

また、各種サービスシステムの登録は、一部のブリーダーを除く場合もあり、ペットショップ、保険代理店、取扱店で登録できる。

⑵　行政（役所・動物愛護協会・動物園等）現在飼っている世帯等が後からでも各種サービスシステムの登録できる様にする為。行政は登録のみ。

⑶　各種サービスシステムの登録は、総体的に管理し情報を調整・交換し、提供していくのを目的とする。

このシステムの前進は、ＡＰＩＣＯ(animals pets insurance company)（対動物サービスシステム及び提供方法)であり､これと平行しているＡＨＳＳＰ(Animal &Human service system　Password)（動物＆人間サービスシステム パスワード）である。これらに保険を絡める構成であり、登録証としての発行は、オールマイティーに使用できるＩＤカード、クレジットカード、マネーカード、パスワードなどである。

※因みに、中国では、犬のみに関して、１世帯１匹のみ１万元で登録が義務付けられているとか。

⑷　ＡＰＩＣＯ、ＡＨＳＳＰの構成、こうすることにより考えられ得る効果

a. 独禁法に抵触しなければ、品種により、ある程度の価格統一が図れる。

b.ペット屋を始めブリーダー等に保障を備えた保険、ＡＰＩＣＯ(animals pets insurance company)（対動物サービスシステム及び提供方法）或いは、ＡＨＳＳＰ（Animal &Human service system Password）（動物＆人間サービスシステム パスワード）などに加入することにより、ある意味では身分保証にもなり、「倒産」等も皆無とはいわないがなくなってくる。

私は、これがないから、ペット屋さんが、経営が安定しないのだとも思う。と同時に格差をなくすことができる。

APICO

2，医療

現行のペット医療保険(健康保険)は、あくまで共済保険です。

⑴　ここでは、人間界の社会保険制度をイメージしています。従って、薬価制も導入し、勿論病気等により治療費等を一定化できる。

そしてここで、ＡＰＩＣＯ(animals pets insurance company)（対動物サービスシステム及び提供方法）或いは、ＡＨＳＳＰ（Animal &Human service system Password）（動物＆人間サービスシステム パスワード）が保険証であり、診察券になり得る。

⑵　ＡＰＩＣＯ、ＡＨＳＳＰの構成、こうすることにより考えられ得る医療効果

a. 現行の共済保険による、医療費の高額化を軽減できる。

b. 薬価制にする事により、自由価格である為の薬代の高額格差を無くす事ができる。

c. 獣医さんがＡＰＩＣＯ(animals pets insurance company)（対動物サービスシステム及び提供方法）或いは、ＡＨＳＳＰ（Animal &Human service system Password)（動物＆人間サービスシステム パスワード）に加入する事により、格差もなくなり、身分保障に繋がる。

※私は、医師、歯科医、薬剤師等が身分保障及び地位が築かれているのは、社会保険制度の存在もあると思っております。

APICO

3，サービス業

⑴　ペットの葬儀屋

道路等でぺちゃんこになった遺体の処理を、今まで、保健所等行政が行っていたものを、これらの民間業者にも委託。

APICO

⑵　トリマー、ペット美容室、医院

トリマーの仕事により、毛をカットするトリミング、シャンプー、爪の手入れ、耳掃除、ペットの指圧・マッサージなどが受けられ、また、医院でもトリマーを行うところもあり、治療が必要なペットまたは高齢である場合は、検査や治療後のトリミングであるが、混合ワクチン未接種のペットは伝染病防止の為にトリミングが出来ない場合もあるからこのよう場合にＡＰＩＣＯanimals pets insurance company）（対動物サービスシステム及び提供方法）或いは、ＡＨＳＳＰ（Animal &Human service system Password）（動物＆人間サービスシステム パスワード）は効果を発揮する。

APICO

⑶　ペットホテル

ペットと一緒に同室で泊まれるホテルの場合、またはペットを預ける場合も、このＡＰＩＣＯ(animals pets insurance company)（対動物サービスシステ及び提供方法）或いは、ＡＨＳＳＰ（Animal &Human service system Password）（動物＆人間サービスシステム パスワー）は効果を発揮する。

APICO

⑷　ペットのリサイクル(飼えなくなったペット等を一旦預かり、里親を斡旋する)等々、これらにもＡＰＩＣＯ(animals pets insurance company) (対動物サービスシステム及び提供方法) 或いは、ＡＨＳＳＰ (Animal &Human service system　Password) (動物＆人間サービスシステム パスワード) により発行で適応できる。

APICO

保 険
Insurance

⑸　ＡＰＩＣＯ、ＡＨＳＳＰの構成 こうすることにより考えられ得るサービス効果

a. 専ら、瑕疵・過誤の保障、シャンプーで犬がカブレた等、ペットに噛まれて怪我をした等の業者への補償。

保険
Insurance

４，ペット葬祭・火葬・葬儀・供養・納骨堂・ペット霊園等の関連サービス

火葬料金はペットの種類により異なり、例えば、小鳥、猫、ウサギ、小型犬、中型犬、大型犬、超大型犬等の合同・個別・立ち会い・訪問火葬があり、これらのことが全て、ＡＰＩＣＯ(animals pets insurance company)（対動物サービスシステム及び提供方法）或いは、ＡＨＳＳＰ（Animal ＆Human service system Password）（動物＆人間サービスシステム パスワード）で決済できる。

５，狂牛病、鳥インフルエンザ

⑴　犬・猫の狂牛病

飼料を介して、または移植により感染するプリオン病である。感染源は羊（スクレイピー病）や牛（BSE）の畜産副産物を含んだペットフードと言われている。感染した犬の症状として「犬が音に対して、異常に吠えて興奮する。怒りっぽくなり、異常に吠えて攻撃的になる。また、攻撃的で噛み付く行動が見られ、認知症（痴呆症）が進行し、転倒したり、起立、歩行が出来なくなる。」

これらのプリオン病の医療を円滑に行うのが、このシステムです。

APICO

⑵　犬・猫の鳥インフルエンザ

犬や猫に鳥インフルエンザウイルスが感染する報道が報告された今日、単なる犬・猫の扱いでなく、家族の一員とされるようになった。このために人にも感染するのでないかと心配し、獣医と医師の相談も必要になった。これに伴なう医療費の支払いにＡＰＩＣＯ(animals pets insurance company)（対物サービスシステム及び提供方法）或いは、ＡＨＳＳＰ（Animal &Human service system　Password）（動物＆人間サービスシステム パスワード）が便利である。

APICO

保険
Insurance

⑶　鳥獣による農作物への被害等による業者への保障

鳥獣による農作物被害については、今迄、損保では対応し得なかった内容であった。これをスムーズに対応できるのがＡＰＩＣＯ(animals pets insurance company)（対物サービスシステム及び提供方法）或いは、ＡＨＳＳＰ（Animal &Human service system　Password）（動物＆人間サービスシステム パスワード）である。

６，ＧＰＳ等による、ペット（動物）の不明。所在の管理

家族の一員でもあるペットが、いなくなった時にＧＰＳ端末が付いていれば、何処にいるか容易に検索で探すことができる。また、リアルタイムで行動軌跡が検索確認できる。

このＡＨＳＳＰ（Animal &Human service system　Password）（動物＆人間サービスシステム パスワード）では、ペット用の小型ＧＰＳの端末と人間用のＧＰＳの端末がセットで契約でき、人間用のＧＰＳの端末では緊急事態が発生時のボタンがセットされ、家族・管理会社・警備会社などに通知される。

⑴　宣伝表現

ペットと人間のＧＰＳセット保険

ペットと人間のＧＰＳセット保険で割安

ペットと人間のＧＰＳセット保険で緊急事態安心

The usage, expressional English

1,

Registration of various service systems

A registration fee changes with each of an animal, or kinds, and a compensation premium etc. is contained.

(1)

In a part of administration, although carried out by limiting to a dog, a window is limited and it is also that the quality is guaranteed by carrying out in a pet store at the time of purchase.

Moreover, registration of various service systems can be registered except for some breeders in a pet shop, an insurance agent, and an agency.

(2)

In order that the household kept now [administration] (a public office, the Society for the Prevention of Cruelty to Animals, zoo, etc.) may enable it to register various service systems even afterwards.

Only for registration, administration is.

(3)

It manages on the whole, information is adjusted and exchanged, and

registration of various service systems is the purpose of offer.

Advance of this system

ＡＰＩＣＯ(animals pets insurance company)

It is parallel to this.

ＡＨＳＳＰ (Animal &Human service system Password)

The registration certificates of insurance are the ID card which can be used

for the Almighty, a credit card, a money card, a password, etc.

Or [incidentally a duty of registration is imposed only for one household

/ one / for 10,000 yuan only about the dog in China]

(4)

The composition of APICO and AHSSP, the effect which may be considered by carrying out like this

a

If it is not against the Antimonopoly Law, a certain amount of price unification can be aimed at by the kind.

b.

By joining insurance, APICO, or AHSS etc. which equipped the breeder etc. with security including the pet store, in a sense, it also becomes a social position guarantee, and although it is not said that there is "no bankruptcy" etc., it is lost.

Since I do not have this, the pet store regards me as management not being stabilized.

Simultaneously, a gap can be lost.

2,

Medical treatment

(1)

Here, the social insurance system of a human community is imagined and introduction of a price-of-medicine system and sick medical treatment expense are fixed.

And the following is an insurance certificate and may become a patient's registration card here.

ＡＰＩＣＯ(animals pets insurance company)

ＡＨＳＳＰ (Animal &Human service system Password)

(2)

The composition of APICO and AHSSP, the medical effect which may be considered by carrying out like this

a.

The large sum of medical expenses by the present co-insurance is mitigable.

b.

By making it a price-of-medicine system, the large sum gap of the charge for medicine for being a free price can be lost.薬価制にする事により、自由

c.

When the veterinarian joins APICO or AHSSP, a gap is also lost and it leads to social position security.

A doctor, dentist, the pharmacist, etc. regard me as that social position security and the status are built having existence of a social insurance system.

APICO

3,

Service industry

(1)

A pet's mortician

Processing of a pet's dead body is also commissioned to the private sector contractor instead of administration.

APICO

(2)

A trimmer, a pet cosmetics room, a hospital

There is also a place which the trimming which cuts hair, a shampoo, care of a nail, ear cleaning, acupressure, a massage of a pet, etc. can be received, and performs a trimmer by a trimmer' s work also in a hospital, and although it is the trimming after inspection or medical treatment when medical treatment is a required pet or advanced age, since mixed vaccine a non-inoculated pet may not be able to do trimming for infectious disease prevention, in a such case, APICO or AHSSP demonstrates an effect.

(3)

Pet hotel

When depositing the case of the hotel at which it can stay in the same room together with a pet, or a pet, this APICO or AHSSP demonstrates an effect.

(4)

A recommendation of the owner of the pet it became impossible to keep can be adapted by APICO or AHSSP.

APICO

(5)

Composition of APICO and AHSSP　The service effect which may be considered by carrying out like this

4,

Related service of pet funerals and festivals, cremation, a funeral, mass for the dead, a charnel, etc.

APICO

5,

Mad cow disease, bird influenza

(1)

Mad cow disease of a dog and a cat

It is the prion disease infected by transplant through feed.

The source of infection is called pet food containing the stock raising by-product of the sheep or a cow (BSE).

As the condition of the infected dog ″to sound, a dog barks unusually and is excited.

It becomes touchy, and it barks unusually and becomes offensive.

Moreover, it is offensive and the action which bites is seen, and (Alzheimer's disease) advances, and or standing up and a walk become impossible. ″ This system performs medical treatment of these prion diseases smoothly.

(2)

Bird influenza of a dog and a cat

It came to consider as a family's member instead of the treatment of mere dog and cat today when the report in which a bird influenza virus is infected with a dog or a cat was reported.

For this reason, it worried about whether it is infected also with people, and consultation of a veterinarian and a doctor is also needed.

APICO or AHSSP is convenient for payment of the medical expenses accompanying this.

(3)

Security to the contractor by the damage to the agricultural products by the wildlife etc.

They were the contents which could not respond by insurance about the agricultural-products damage by the wildlife till now.

APICO or AHSSP can respond this smoothly.

6,

The pet (animal) by GPS etc. is unknown.

Management of the whereabouts

If the GPS terminal is attached when the pet which is also a family's member stops there being, it can be where or can search by reference easily.

Moreover, the reference check of the action locus can be carried out on real time.

At this AHSSP, the terminal of small GPS for pets and the terminal of GPS for man can contract by the set, at the terminal of GPS for man, the button at the time of generating is set and a family, a commissioned company, a security company, etc. are notified of state of emergency.

(1)

Advertisement expression

GPS set insurance of a pet and human being

Relatively cheap by the GPS set insurance of a pet and human being

It is state-of-emergency relief at the GPS set insurance of a pet and human being.

あとがき

主旨は:少子化に伴い、ペットを含めた動物との共存の社会生活になって来ております。しかし、それに伴い、ペットを含む動物との関連事業者個々の連携や、取り巻く環境の整備が何一つできておりません。それら全てを、保険を絡める事により、時としては支援し、またすべての関連情報の提供窓口でもある機構というか、システムであり、このシステムの前進は、ＡＰＩＣＯ(animals pets insurance company)（対動物　サービスシステム及び提供方法）であり、これと平行しているＡＨＳＳＰ（Animal ＆Human service system　Password）（動物＆人間サービスシステム パスワー）である。これらがこれからの時代に適応したシステムである。

ペットと一緒に旅先のホテルで泊まれる建物として関連するのが「車庫付き中・高層建築物等における車等の安全昇降機」【ISBN】　978-4-938480-95-0 2015/11/02発行で著者が解説している。この車庫付き中・高層建築物は各階の住居に駐車場・車庫を設けた特長の使用・表現を解説し、更に車両のエレベーター「回動エレベーター、ハイパーエレベーター、シリンダーエレベーター、シリンダー回動エレベーター」等で1階から車庫付住居、オフィスなどに昇降するものである。これらの構成・使用・効果などの表現を創作した。また、本件の各エレベーターのメンテナンスマニュアル、製造工程のマニュアルなども別途にある。

牧野真一

動物＆人間サービスシステム　パスワード&IDカード

定価（本体 1,500 円＋税）

２０１５年（平成２７年）１２月２４日発行

No. 
MN96-7-013

発行所　IDF(INVENTION DEVLOPMENT FEDERATION)
発明開発連合会®
ﾒｰﾙ 03-3498@idf-0751.com　www.idf-0751.com
電話 03-3498-0751㈹
150-8691 渋谷郵便局私書箱第２５８号
発行人　ましば寿一
著作権企画　IDF 発明開発(連)
Printed in Japan
著者　牧野 真一 ©
(まきのしんいち)
